# Tim

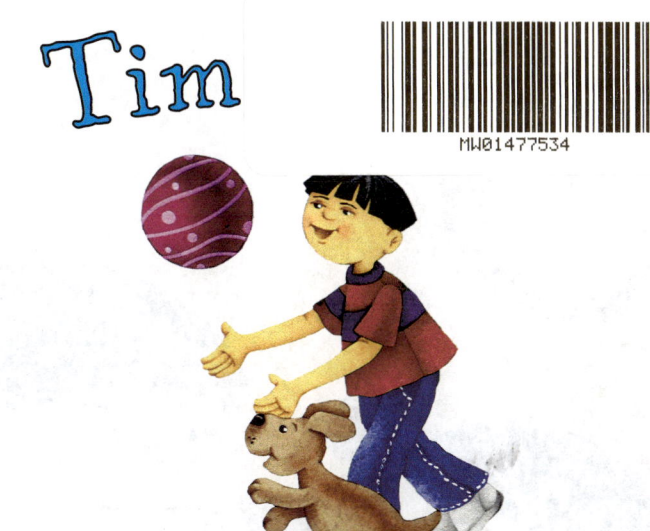

by Suki Sataka
illustrated by Ana Ochoa

**HMH**

Copyright © by Houghton Mifflin Harcourt Publishing Company

All rights reserved. No part of this work may be reproduced or transmitted in any form or by any means, electronic or mechanical, including photocopying or recording, or by any information storage and retrieval system, without the prior written permission of the copyright owner unless such copying is expressly permitted by federal copyright law. Requests for permission to make copies of any part of the work should be submitted through our Permissions website at https://customercare.hmhco.com/contactus/Permissions.html or mailed to Houghton Mifflin Harcourt Publishing Company, Attn: Intellectual Property Licensing, 9400 Southpark Center Loop, Orlando, Florida 32819-8647.

Printed in the U.S.A.

ISBN 978-1-328-77230-5

4 5 6 7 8 9 10  2562  25 24 23 22 21

4500844736         A B C D E F G

If you have received these materials as examination copies free of charge, Houghton Mifflin Harcourt Publishing Company retains title to the materials and they may not be resold. Resale of examination copies is strictly prohibited.

Possession of this publication in print format does not entitle users to convert this publication, or any portion of it, into electronic format.

Mark and Mary can play after they eat lunch.
Lunch is at 12 o'clock.

Mary and Mark feed Fuzzy.
Fuzzy always eats an hour
before they do.

What time does Fuzzy eat?

Mary feeds her bird.
Chirpo eats a half hour before Mark and Mary.

Did Mary feed her bird before or after feeding Fuzzy?

Bubbles is hungry, too.
Mark feeds his fish a half hour before lunch.

What time is it? 5

Finally, Mark and Mary eat lunch.

6   What time is it now?

Playtime!

# Responding

**Vocabulary**

## A Clock and a Clock

**Draw**

Look at pages 4 and 5. Draw the two clocks you see.

**Tell About**

Compare and Contrast Look at the clocks on pages 4 and 5. Tell how the clocks are different. Tell what time each clock shows. Tell how the clocks are alike.

**Write**

Look at pages 4 and 5. Write the time each clock shows.